UNTAINS

Printed in China

02 03 04 05 06 5 4 3 2 1

Library of Congress Cataloging-in-Publication Data
Price, Martin F.
Mountains : geology, natural history & ecosystems / Martin F. Price.
p. cm. — (Worldlife library.)
Summary: Introduces the formation, ecology, diversity, and importance of mountains.
Includes bibliographical references (p. 72).
ISBN 0-89658-251-5 (pbk. : alk. paper)
1. Mountains—Juvenile literature. [1. Mountains. 2. Mountain ecology.
3. Ecology.] I. Title. II. World life library.
GB512 .P75 2002
577.5'3—dc21

2002006155

Distributed in Canada by Raincoast Books, 9050 Shaughnessy Street, Vancouver, B.C. V6P 6E5
Published by Voyageur Press, Inc.
123 North Second Street, P.O. Box 338, Stillwater, MN 55082 U.S.A.
651-430-2210, fax 651-430-2211
books@voyageurpress.com   www.voyageurpress.com

Educators, fundraisers, premium and gift buyers, publicists, and marketing managers:
Looking for creative products and new sales ideas? Voyageur Press books are available at special discounts when purchased in quantities, and special editions can be created to your specifications. For details contact the marketing department at 800-888-9653.

MOUNTAINS

Martin F. Price

**WORLDLIFE
LIBRARY**

Voyageur Press

The Grand Tetons at sunrise, Grand Teton National Park, Wyoming, U.S.A.

Contents

Introduction   7

Why Do Mountains Matter?   11

What and Where are Mountains?   13

Mountains Are Not Eternal!   17

Water Towers for the World   25

Vertical Mosaics   33

Centers of Diversity   47

Places of Attraction, Challenge, and Renewal   55

Safeguarding Our Mountain Heritage   63

Recommended Reading, Web Links and Information   71

Index/Biographical Note   72

Introduction

Ascend a mountain, feel the freshness of the air, bathe in the chilly water of a mountain river, look at those enormous stones which are much older than you, and harmony will settle your soul.

Maria Barannikova, 'Thoughts on mountains'. In Para Limbu (ed),
Mountains Forever. International Centre for Integrated Mountain
Development (ICIMOD), Kathmandu, 2001

Mountains have been an important part of my life since my parents first took me to Snowdonia at the age of five. The Welsh hills seemed far away from my home in London, but I was immediately captivated by the challenge of climbing up parts of them, and I remember the different colors: many greens, and the purples and grays of heather, slate, and sky. A year later, we went to the Swiss Alps, and I found myself surrounded by the translucent green ice of the Morteratsch glacier. From then until I was about 20, my main interest in mountains was to climb them. At 16, I climbed the Jungfrau in Switzerland, together with three Yugoslav climbers who were glad to find a fourth person to climb with. They shared their pickled vegetables with me; an early instance of the generosity of mountain people and my feeling that food always tastes better in the mountains. Nearly 30 years later, I was back at the same place, sharing a very different lunch with the former President of Switzerland during an evaluation of the area as a potential World Heritage Site, followed by a fantastic helicopter ride with him back to Interlaken. The mountains hadn't changed much – though the glaciers were smaller – but I saw them in a new way.

In 1978, I left Europe and spent most of the next 13 years living on the east slopes of the Rocky Mountains of Alberta and Colorado. Mountains were now my professional interest, and I came to realize that finding solutions to most 'environmental problems' in the mountains really means understanding how people use and perceive them, and the limitations of their very diverse environments. The Canadian Rockies were very different in scale from European mountains, with a more expansive beauty punctuated by lakes of many different hues of blue. Yet though I loved the spaciousness and the clear air, something was missing: people who lived in the mountains. In Colorado,

Crib y Ddysgl, a peak in the Snowdon range. Snowdon is the highest mountain in Wales, U.K.

I found more people in the mountains, but missed the landscapes of Europe's mountains, shaped by centuries or more of management.

Returning to Europe in the 1990s, I began exploring my family history in the countries emerging from communism, visited many of their mountains, so close to western Europe, but largely unknown, and worked with people living in them to widen our knowledge of their richness. I have very fond memories of the generosity of local people in Bulgaria, the Caucasus, Serbia, and many other places. Furthest east, in the Altai, I found hillsides redolent with the smells of many wild herbs, and ate in the yurts of herders from Kazakhstan. In the Caucasus, I walked through some of the few remaining primeval mountain forests in Europe, among huge trees both standing and fallen. But my main interest had become less the natural features of the mountains than the people who live in these challenging environments, finding ways to live through cultures which have evolved to make the best of all the opportunities, but often ignored or inadequately compensated for all that they provide to us.

The Matterhorn, Swiss Alps, surrounded by shrinking glaciers.

Mountains can be challenges in many ways, but what makes them special are the feasts which they, and their people, provide for all our senses: a heritage to foster.

Two Jack Lake and Mount Rundle in Banff National Park, the oldest in Canada, beginning with the designation of the area around its hot springs in 1885.

Why Do Mountains Matter?

Wherever we may come from, however high or small the hills or mountains may be in the land of our birth, we are all mountain people. We are all dependent on mountains, connected to them, and affected by them, in ways we may never have previously imagined.

Jacques Diouf, Director-General, Food and Agriculture Organization of the United Nations.

Mountains cover a quarter of the Earth's land surface. They provide water for at least half of the world's people. They are global centers of cultural, biological and geological diversity. And they are visited by hundreds of millions of tourists every year. Yet, until quite recently, the world's many mountain regions seemed to have little importance at the global scale. Most of the countries which are mainly mountainous are minor players on the world stage. Equally, the mountain areas of major countries, even those inhabited by hundreds of millions of people, such as China and India, have generally been seen as of marginal importance by politicians and policy-makers living in lowland capital cities.

This picture has begun to change. In the months leading up to the Earth Summit, held in Rio de Janeiro in 1992, a small group of scientists and development experts who had worked in mountain regions around the world recognized that this meeting would provide a unique opportunity to alert the world to the diverse values of mountains. Calling themselves 'Mountain Agenda', they managed to get a chapter on mountains into 'Agenda 21', the plan for action for the twenty-first century that came out of Rio. This effectively put mountains on the global agenda, with the other 'big issues' such as climate change, tropical deforestation, and desertification.

Over the next few years, representatives from governments in most parts of the world met to consider the ways in which mountains were important to their countries. Some governments created mountain institutions or passed laws to benefit mountain people. New networks of mountain people and scientists emerged. And, led by the government of the Central Asian state of Kyrgyzstan, a further momentum built, to declare an International Year of Mountains. In 1998, the U.N. General Assembly declared that this would be in 2002, providing a unique opportunity to raise awareness of the diverse values of mountains to everyone: as the slogan for the Year states, 'We are all mountain people'.

Cuernos del Paine, Torres del Paine, Chile.

Mountains cover about half of the area of Antarctica, making it the most mountainous continent on Earth.
These uninhabited mountains are visited by only a fraction of the 14,000 people who travel to Antarctica each year.

What and Where are Mountains?

Mountains are big. Very big. But they are also great. Very great. They have dignity and other aspects of greatness.
Arne Naess, 'Modesty and the conquest of mountains'.
In M.C. Tobias and Harold Drasdo (eds) *The Mountain Spirit*, 1979

Everyone who has seen or, better still, walked or climbed up a mountain knows what it is. Mountains are steep, and there is a significant difference in altitude from the bottom to the top. Yet although we know a mountain when we see it, there are many definitions of what mountains are, usually based on altitude; but there are parts of the Earth which are at high altitude but which are very flat; these include the Tibetan Plateau, the altiplano of the Andes, the high plains of western North America, and the plateaux of Eastern Africa. In Europe, governments have defined the parts of their countries above a specific altitude as 'mountainous', recognizing that conditions for agriculture generally get worse as altitude increases, and so farmers at higher altitudes need financial support to survive. Any land above 1970 ft (600 m) in Bulgaria or Slovenia is 'mountain'; in Austria, the threshold is 2950 ft (900 m); and in Spain, 3280 ft (1000 m). So, defining mountains is not a simple task, and often has political implications.

In the 1990s, the U.S. Geological Survey developed a database recording the average altitude of every square mile of the Earth's surface. This provided the starting point for the first effort to create a global mountain map according to a set of rules. Criteria relating to altitude, slope, and topography were developed, and scientists, mountaineers, and policy-makers were asked to evaluate different combinations of these criteria for the mountain areas they knew best. In this study, it was relatively easy to state that all parts of the Earth more than 8200 ft (2500 m) above sea level are mountains because of the physiological challenges to living there, and to 'filter out' high-altitude plateaux and plains by setting an appropriate slope threshold. However, the unique element of this work was that it considered the roughness of the topography – only possible with such a detailed, global database. The agreed criterion was that, for a place to be defined as mountainous, the altitude had to vary by at least 985 ft (300 m) over a radius of 4 miles (7 km). The resulting map thus showed all of the older and lower mountain systems, such as the Appalachians, the Scottish Highlands, the Atlantic Highlands of Brazil, the Urals, the Australian Alps, and the mountains of southeastern China.

Mountains of the World

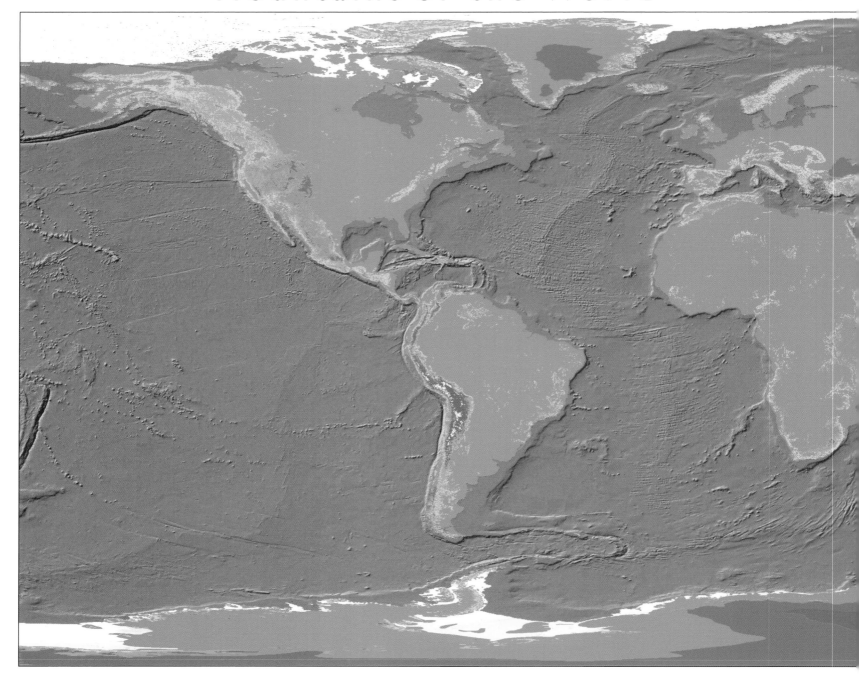

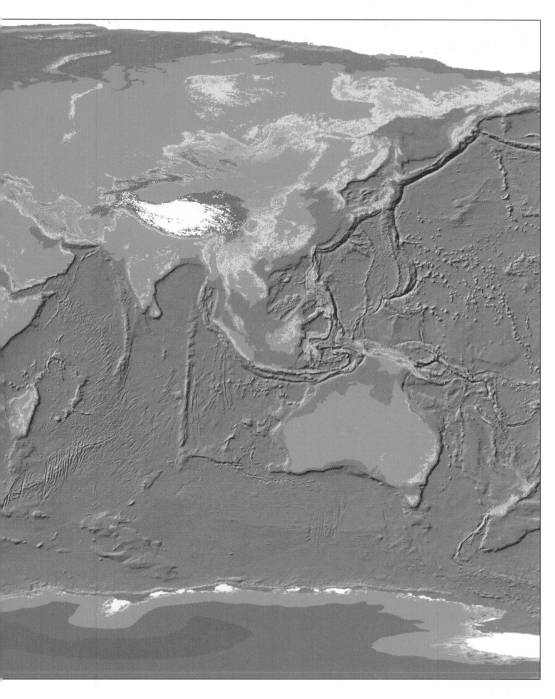

The first map of the world's mountains, defined according to a consistent set of criteria relating to altitude, slope and topography (see page 13). It shows that mountains cover 13.82 million sq miles (35.8 million sq km): 24 per cent of the Earth's land surface, on every continent, and from the Equator almost to both poles.

Mountain Classes

Elevation > 4500m

Elevation 3500 - 4500m

Elevation 2500 - 3500m

Elevation 1500 - 2500m, and slope > 2°

Elevation 1000 - 1500m, and slope > 5° or local elevation range > 300m

Elevation 300 - 1000m, and local elevation range > 300m

Non-mountainous land

UNEP WCMC

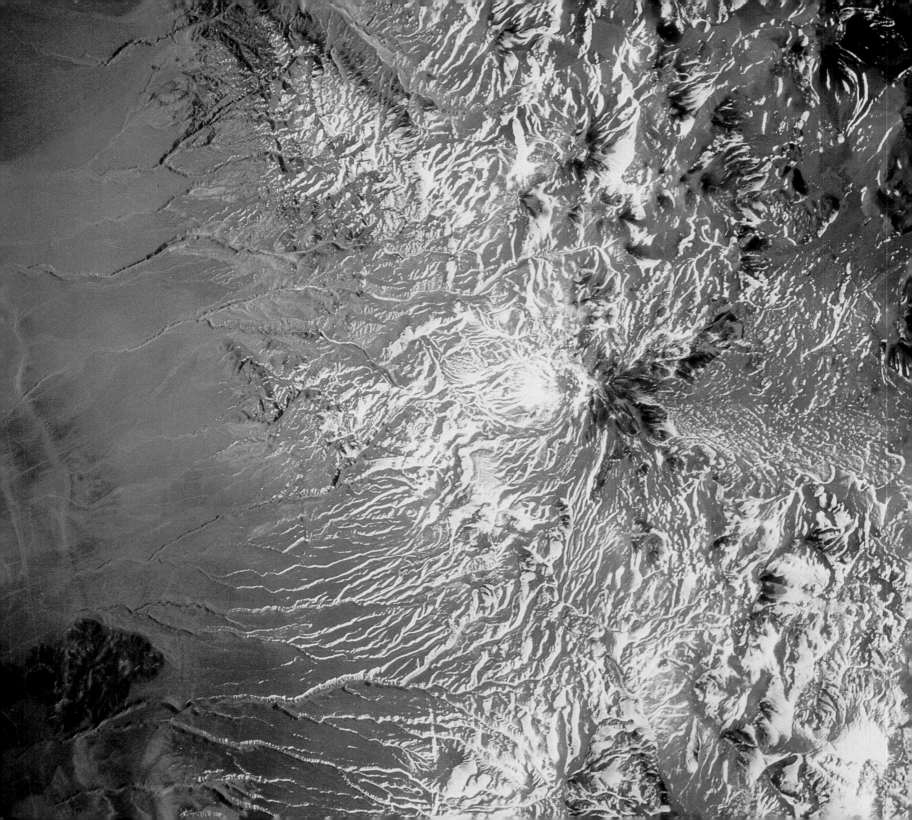

Mountains Are Not Eternal!

Every valley shall be exalted, and every mountain and hill made low. Isaiah XI: 4

On the timescale of human lives, most mountains seem eternal. It is only rarely that a new volcano emerges, or that part of a mountain disappears as a result of volcanic activity, such as Mount St. Helens in the Cascades of the United States in 1980, or Mount Pinatubo in the Philippines in 1991. Yet on geological timescales, mountains come and go: the low mountains of the Appalachians and the Scottish Highlands may once have been as high as today's highest mountain range, the Hindu Kush-Himalaya.

The main process leading to the formation of mountains is the collision of the lithospheric plates that form the outer 'skin' of our planet. As the plates move slowly together, the sediments above them are compressed and uplifted, leading to folding and faulting. There have been three major mountain-building periods, or orogenies. The Caledonian orogeny took place about 550-370 million years ago, resulting in the mountains whose last remnants are now found in Scotland, Scandinavia, Greenland and the northern Appalachians of eastern North America. During the Hercynian, Variscan, or Appalachian orogeny, about 200-400 million years ago, North Africa collided with Europe, producing what are now the low or 'middle' mountains of the Massif Central in France, the Vosges on the French/German border, the Black Forest in Germany, and the Giant Mountains along the Czech/Polish border. Over the same period, continental collisions produced the southern Appalachians in the United States.

The Alpine orogeny started about 65 million years ago, and the mountains that began to be created then are still growing. These include the Alps, Himalaya, and the Andes, the mountains of western North America, and other ranges around the Pacific Rim. The fastest growing are the Himalaya: recent satellite measurements show that they are rising at 2.3 in (6 cm) a year; the Andes are growing at only about 3/10 of an inch (4 to 5mm) per thousand years.

Mountains are also formed by volcanic activity. Most are along the margins of plates, particularly the 'Ring of Fire' around the Pacific Ocean. Others are above isolated 'hot spots' in the Earth's crust, for instance Italy's Mount Etna and the mountains of Hawaii. In fact, Mauna Loa, Hawaii's highest peak, is

The Potosi region of the Bolivian Andes from space, with the high altiplano to the left.

the tallest on Earth: 29,856 ft (9100 m) when measured from its base on the sea bed to its summit 13,681 ft (4170 m) above sea level. Shield volcanoes, such as those of Hawaii, have relatively gentle slopes because they are made of runny basaltic lava. Other types are explosive, producing gas, ash, and thick andesitic lava which forms steep cone-shaped volcanoes, such as Japan's Mount Fuji. However, not all volcanic mountains are on land. In the middle of the Atlantic Ocean, upwelling magma has created a range of underwater mountains 9300 miles (15,000 km) long. Part of this rises 13,000 ft (4000 m) above the ocean floor, forming the mountains of Iceland and islands such as the Azores and St. Helena.

Wearing Down Mountains

The folding, faulting, and volcanism that create mountains are only the first part of the story of how they are formed and sculpted. As soon as mountains begin to grow, geomorphological processes, working together with gravity, begin to wear them down. These processes operate at many scales in both space and time. The smallest scale is that of weathering: the gradual loss of rock surfaces due to rapid variations in temperature, frost shattering, and chemical changes. A typical characteristic of mountain areas is major changes in temperature that occur over each 24-hour period, leading to insolation weathering, in which fragments of rock break off after being heated by the sun to high temperatures during the day, and then cooling at night. These temperature changes often also mean that any moisture — whether rain or melted snow — freezes during the night, breaking off rock fragments in the process of frost weathering. Chemical weathering, the solution of rock by naturally occurring chemicals in this water, also gradually removes rock, more rapidly in tropical areas and near sea coasts. Although all of these processes operate at very small scales, their combined result, over years, centuries, or millennia, is to wear mountains away, creating screes and rock debris.

On a very different timescale, glaciers have been a major force of change in many mountain areas. Moving gradually downhill under the influence of gravity, glaciers erode mountains by plucking rock from rock faces, and grinding and scouring rock surfaces to create a range of typical features, such as U-shaped and hanging valleys, pyramidal peaks, and sharp ridges, which are then maintained by ongoing weathering

Waimea Canyon, Kauai, Hawaii, formed by water erosion.

processes. Glaciers also act as conveyors, moving the debris that has fallen on to them as a result of weathering and rockfalls. Some of the rock debris, or till, that results is left as moraines — long ridges of loose rock that show the past limits of glaciers, sometimes blocking valleys and forming lakes. Meltwater rivers flowing beneath the ice also transport large volumes of rock debris. Where these rivers emerge at the front of the glacier, this is deposited as outwash, which infills the valley bottom. Usually, a large proportion of the till flows downstream, in rivers that have a characteristic milky color.

For the last century or so, most of the world's glaciers have been shrinking as a result of climate change. During the Pleistocene era, about 1,800,000 to 11,000 years ago, many of the world's mountains were far more glaciated than now, with much greater rates of erosion than achieved by today's rivers. In the Appalachians the average rate of erosion during the last Ice Age was 3 to 10 ft (1 to 3 m) per 1000 years; current rates of erosion by rivers are 3/4 to 1 in (20-30 mm) per 1000 years. Yet the large amount of water present in most mountain areas means that streams and rivers are important in removing the products of weathering and glacial erosion and, more gradually, the sediments and rock across which they flow. Rates of fluvial erosion vary considerably over time: often, a very large proportion is removed in a rather small number of floods, which occur after major storms or at the beginning of spring, when the winter's snow melts rapidly. Water erosion and deposition lead to many of the typical features of mountain landscapes, such as narrow V-shaped valleys, waterfalls, terraces, alluvial fans and wide flood plains.

Hazardous Landscapes

In addition to the gradual processes of weathering and glacial and water erosion, parts of the mountain landscape are also modified by 'natural hazards' such as avalanches, rockfalls, and landslides — often partly caused by human activities. While avalanches primarily damage trees and shrubs in their path — and sometimes houses, roads, railways, and tourist facilities — the largest rockfalls and landslides can leave major lasting marks on the landscape. Earthquakes and volcanic eruptions can affect even larger areas. Combined, such processes can have devastating effects. When an earthquake resulted in the

Mount St. Helens in July 1980, two months after the main eruption which removed 3 cu km of the mountain.

collapse of the ice and rock near the summit of Huascaran in Peru on 31 May 1970, the subsequent mudflow, moving at 300 miles (480 km) an hour, devastated the town of Yungay, 7 miles (12 km) away, killing 18,000 people in the town, causing 52,000 further deaths, and leaving 200,000 people homeless.

The Huascaran catastrophe resulted from a highly unfortunate linked series of events, and was largely unpredictable. In a warming world, another type of mountain catastrophe may well be more frequent: glacial lake outburst floods (GLOFs). Although glaciers are made of ice, lakes can form on them as a result of melting by the sun, and also beneath them, where meltwater rivers become restricted. Lakes form also behind the terminal (front) moraine of glaciers. GLOFs occur when the lakes overflow, or when surface water melts through to the base of the glacier. Immense volumes of water are suddenly released, often causing devastating damage and many deaths. At least 12 GLOFs have been recorded since 1935 in Tibet, where the 1954 GLOF from Sangwang Cho glacial lake at the headwaters of the Nyangqu river released over 392 million cu. yd (300 million cu. m) of water, creating a 130 ft (40 m) high wave. The flood deposited up to 15 ft (4.5 m) of debris in the upper valley and caused damage in the cities of Gyangze and Xigaze, 75 and 125 miles (120 and 200 km) downstream.

GLOFs have also caused significant damage and deaths in other parts of the Himalaya, the Andes, and the Alps over the past century. By constructing inventories of glaciers and glacial lakes, it is possible to establish which lakes form a threat, and to install early warning devices and mitigation measures to release the water gradually. In Nepal, these have been installed on the Tsho Rolpa lake, recognizing that a GLOF from this lake could cause damage up to 60 miles (100 km) downstream, threatening over 10,000 people and much infrastructure, including a hydropower plant.

Mountain people who have a long experience of a mountain area understand the likelihoods of natural hazards, and avoid dangerous areas, or take measures to control hazards, but such knowledge is not always used or recognized. While our ability to predict the timing of earthquakes and volcanic eruptions is improving, it is far harder to predict where the impacts will occur. In the 1991 eruption of Mount Pinatubo in the Philippines, although fewer than 400 people were killed, the eruption affected over 400,000 people directly, 1.5 million indirectly, and influenced the world's climate for three years.

Gilkey Glacier, Alaska, U.S.A., showing the bands of debris it has carried down-river.

Cotopaxi (19,388 ft, 5911 m) in Ecuador is the world's highest active volcano, and has erupted 50 times since 1738. The 1877 eruption melted snow and ice on the summit, producing mudflows that traveled 60 miles (100 km). The shrinkage of its glaciers may lead to future disasters.

Water Towers for the World

We complain of the mountains as rubbish, as not only disfiguring the face of the earth, but also to us useless and inconvenient; and yet, without these, neither rivers nor fountains nor the weather producing and ripening fruits could regularly be produced.

Archbishop William King, *On the Origin of Evil* (1731)

Mountains are often described as water towers – the sources of freshwater for billions of people. All of the world's major rivers originate in the mountains. Between a third and a half of all freshwater flows come from mountain areas; more than half of humanity relies on mountain water for drinking, domestic use, fisheries, irrigation, hydro-electricity, industry, recreation, or transportation. While mountain areas occupy only relatively small proportions of most river basins, the reason that a large proportion of the precipitation – rain, snow, etc. – falls in them is simple: as air rises over the mountains, it cools, releasing the moisture it holds. The greater height of the mountains is important not only for triggering precipitation, but also because temperature decreases with altitude. This means that there is less evaporation once the precipitation has fallen, and also that it is more likely to fall as snow than as water – one reason that many mountains have names which mean, or include, the local word for 'white'.

For people living in the lowlands below, the storage of winter precipitation as snow or ice is especially crucial, because when temperatures rise in the spring and summer, this snow and ice melts. The water that is released enters the rivers, flowing downstream to the lowlands, sometimes thousands of miles away, at a time when it is most needed for irrigation and other uses. In the dry parts of the world, mountains are often the only areas which receive enough precipitation to generate runoff and recharge groundwater, typically providing 70 to 95 per cent of the flow to nearby lowlands. For example, the watersheds of the Blue Nile and Atbarah rivers, which rise in the Ethiopian Highlands, occupy only 10 per cent of the Nile river basin, but contribute 53 per cent of the annual inflow to Lake Nasser, and 90 per cent of the sediment input. The remainder of the inflow comes mainly from the White Nile, which flows from the mountains of East Africa.

It is not only in dry parts of the world that mountain water is essential to life and economies. Even in humid areas, mountain water contributes 30 to 60 per cent of the water flowing to the lowlands. In Europe, although the Alps cover only 23 per cent of the area of the Rhine river basin, they provide half the total flow. Other parts of the Alps form a third of the area of both the Rhône and Po river basins and contribute 47 and 56 per cent, respectively, to the lowland flow.

All of these figures are based on the flows in rivers in recent decades. However, climate change already seems to be leading to changes in patterns of flows, and computer models suggest that, over the coming decades, there will be major changes in the timing of precipitation, and more and more rainfall instead of snow. The melting of glaciers will contribute to increased flows in rivers downstream for a while, but once the ice is gone, these flows will decrease dramatically. These trends, combined with increased evaporation in both the lowlands and the mountains, mean that mountain water will become even more important.

Harvesting Mountain Water

About 70 per cent of the freshwater that is taken from rivers worldwide is used for irrigation, in both mountain and lowland areas. Irrigation systems are found in mountain areas around the world, storing water and directing it to fields at the right time and place to maximize crop yields. The most simple systems involve just blocking streams and allowing the water to flood over meadows, as is done on traditional hay meadows in the Alps. More complex systems involve the construction of channels to bring water from high springs, streams, or glaciers to the fields. These systems can extend for tens of miles, sometimes including channels that have been blasted from rock faces, or are made of planks, suspended around cliffs. The most complex are the underground systems – qanat or khettara – a technology that was probably first developed about 2500 years ago in Iran, then spread eastward to Afghanistan and westward through North Africa. They include water-collection systems, storage reservoirs and cisterns, and underground pipes that carry the water to the fields. The great advantage is that evaporation is minimized, but these systems also require high inputs of labor to build and maintain. Though many, some centuries old, are still in use in the Middle East and North Africa, many others have fallen into disrepair because the required workforce is no longer available, or pumped groundwater is a more easily acquired source of water.

Most glaciers and permanent snowfields, such as those on this peak in Nepal, are shrinking.

A much more recent technology, with very low maintenance costs, is used for harvesting water in some of the driest mountain areas of the world, such as Chile's Atacama Desert. The water contained in the clouds rising over such mountains, especially in the afternoon and at night, does not always condense into rain, especially where there is little vegetation. However, it is possible to 'catch' the water in the clouds by erecting high fences of polypropylene mesh, or 'fog catchers' on which the water condenses. It is then possible to harvest and store this water, and pipe it to villages. Similar projects have been installed in Cape Verde, Ecuador, Namibia, Oman, and Yemen.

Power From Water

The steep gradients of mountain rivers and streams mean that they have great potential for generating energy. The simplest technology is the water mill, developed centuries ago for grinding grain. The great advantage of traditional water mills is that they can be constructed from local materials, and are easily maintained. They are used around the world: in the Himalaya and Hindu Kush, for example, about 200,000 are used to grind grain in mountain villages. Some of these traditional mills have recently been upgraded to provide electricity. The most simple method is to fix a bicycle wheel to the grinding stone, so that as this rotates, the wheel can drive a belt to charge a battery using an alternator. The advantage of such a system is that all of the components are locally available. To produce more substantial amounts of electricity – and also for milling grain – turbines can be installed in streams and rivers. Simple and relatively inexpensive systems are now widely available. The greatest challenge to making them economically viable is to ensure that the electricity, which is generated 24 hours a day, is used not just in the evenings for domestic purposes, but also during the day to provide new economic opportunities – for instance in small-scale industries, such as processing agricultural products and handicrafts, and for telecommunications. In areas with many tourists, the electricity can be used for cooking. This can be important in decreasing the demands on local forests for firewood, a serious impact of tourism in the mountains of many developing countries.

Such small-scale initiatives are one end of the spectrum of hydro-electricity generation in the

The foothills of the Andes in Chile's Atacama desert, one of the driest regions on Earth.

mountains. At the other end are large projects, often with sequences of large dams, such as the 13 on the Columbia River in Canada and the United States, which is one of the most altered rivers in the world. Such projects primarily benefit lowland people, who often gain not only electricity, but also water for irrigation, flood control, and more reliable navigation. Hydropower provides 19 per cent of the world's total electricity supply, in over 150 countries. The global potential for hydropower development is huge, but levels of development vary greatly; while Norway has developed 65 per cent of its potential, Nepal and Ethiopia have developed less than one per cent of theirs.

Sharing the Benefits of Mountain Water

Irrigation systems in mountain valleys are usually constructed by entire communities, as these relatively large projects – at the village level – require major labor inputs and benefit everyone. Once the systems have been constructed, institutions and sets of rules are necessary to ensure that the systems are maintained and the benefits are fairly distributed. Such communal institutions are particularly characteristic of mountain areas, and have also been developed for other communal resources, such as grazing land and forests. It is common for the village council or elders to appoint someone to be responsible for maintaining the irrigation channels and opening the sluices according to a strict rotation, to ensure the fair distribution of water. The rules governing distribution can be quite complicated. When several villages share a system, as in the Hunza valley of Pakistan, each village appoints a guardian for the same purpose. In traditional societies, payment for such services is often in crops and livestock, rather than money. Similarly, water mills are also often constructed communally, and each family has the right to a specified number of hours to grind their grain.

Small-scale hydroelectricity projects are mainly developed for the benefit of local people, who are often adversely affected by large-scale projects, particularly the flooding of valley land, which is usually the most productive for agriculture and also where settlements are located. The need to compensate mountain people for providing downstream benefits was first recognized in a 1916 law in Switzerland, which entitles communities to substantial annual payments and quotas of free energy for granting the rights to hydropower development on their land. Such compensation mechanisms are also being implemented in developing countries. In Brazil, a 1989 law requires that, of the royalties paid for using water for power generation, 45 per cent goes to the affected districts, 45 per cent goes to the affected state(s), and only 10 per cent goes to the federal government. Through such mechanisms, mountain people can directly benefit from the services that their land, and their good management of it, provide to society at large – and some of the compensation can be used to develop new local sources of energy that can improve the quality of life in the mountain communities.

At even larger scales, over 200 river basins, home to 40 per cent of the world's population and covering more than 50 per cent of the global land area, are shared by two or more countries. This often leads to political tension over the distribution of water flows between the states in a river basin – for instance those of the Euphrates, Ganges, Jordan, and Nile – all of which originate in mountains. In Kyrgyzstan, Central Asia, the use of water from the Syr Darya and Amu Darya rivers for generating hydro-electricity has decreased the water available to irrigate agriculture in Kazakhstan and Uzbekistan.

Mountain water irrigates the rice fields in Vietnam.

In turn, diversion of water for irrigation, combined with evaporation and infiltration, has led to conflicts and to the loss of 44 per cent of the volume and 50 per cent of the area of the Aral Sea. Resolution of the resulting tensions by the region's governments is complicated by inadequate measurements of water flows and mutual mistrust.

In the 1990s, the distribution of water from mountains was the cause of many international conflicts. It is likely that there will be more, as populations grow, demands for water increase, and climate change affects the availability of water. Appropriate technologies and new institutions will be needed at all scales, to ensure that the benefits of mountain water are shared as fairly as possible among both the guardians of the water towers and the people living downstream.

Vertical Mosaics

... the huge mural precipices, the deep chasms between the rocks, the waterfalls of unknown height, the hoary remains of the primeval forest, the fields of eternal snow, and the deep black lakes at the foot of the precipices.
John Hill Burton, *The Cairngorm Mountains* (1864)

Alpine vegetation is often like a garden, a mosaic of beauty, a small-scale multitude of ways of coping with life.
Christian Körner, *Alpine Plant Life* (1999)

From a distance, or from the air, many mountains and mountain ranges look like layer cakes, with snow and glaciers icing their summits. Below these come the lightly colored grasslands or tundra, then forests – typically with the darker conifers at higher altitudes, and deciduous trees below – and, in the valleys beneath, houses, roads and railways, settlements and fields. While each of these altitudinal zones is characterized by its vegetation, they can also be described in terms of their climate, soils, and human uses. These different zones occur whether or not people have been using the landscape – but in many mountain regions, the area covered by each zone has been greatly modified by human activities, creating a complex mosaic of vegetation types. In some cases, sustained deforestation, followed by heavy grazing, fire, or both, have led to the loss of most of the vegetation, as on many of the mountains around the Adriatic and Mediterranean Seas from two millennia ago and, more recently, in Haiti and other tropical areas.

This zonation is one aspect of mountains that makes them different from other parts of the Earth. The result of the rapid change in altitude is comparable to a compression of latitudinal zones; for instance, in a mountain region in the tropics, one can travel from tropical rainforest to arctic-like conditions on a glacier in a distance of tens of miles; the equivalent of a journey of thousands of miles on flat land. Yet not all mountains have marked zonation: tundra completely covers the mountains near the Arctic Circle, and very dry mountains have hardly any vegetation at all. There are also many tropical mountains which are entirely covered with forests – though the tree species change with altitude.

One characteristic of many traditional mountain societies is that they use the different zones of their mountains – and sometimes adjacent lowlands – in complementary ways. Transhumance is still

Los Glaciares National Park, Patagonia, Argentina.

practiced by nomadic pastoralists in many parts of the world, including the Balkans, Pyrenees, and Himalaya. They keep their animals in the lowlands during the winter and gradually move them upwards in the spring, reaching the most nutritious high pastures in the height of the summer, when the plants have had time to grow after the winter snows have melted. The process is then reversed in the fall. From the Andes to the Alps to Papua New Guinea, farmers cultivate sometimes very small parcels of land at different altitudes so that they can grow different crop species and varieties at each location, benefiting from the different microclimates. This spreading of effort and risk ensures yields that are sufficient for their needs and possibly some surplus which they can sell.

A great variety of patterns of such vertically arranged land uses exists, implemented by people who are not exclusively 'herders', 'farmers', or 'foresters', but who pursue myriad activities. One typical feature of this system is that the best land for growing crops, usually around permanent settlements, is privately owned. Forests are often communally owned – unless they have been taken over by the state – and the high pastures above are frequently under communal management.

Life Above the Trees

Even on the highest rocky slopes of mountains, lichens often establish themselves. Not far below, mosses and small plants grow in crevices and sheltered places. Although these high places may seem very inhospitable, there is often plenty of sun, which keeps ground-level temperatures high for much of the day, and a good supply of water from rain and snowmelt. The slow-growing plants that live in these areas are well adapted to take advantage of whatever is available during annual growing seasons, often very short, and to survive extreme temperatures, drought, and high winds. The same holds true for a number of hardy insects; springtails have been found above 19,600 ft (6000 m) on snow and ice in the Himalaya, and flies, butterflies, beetles, and spiders have also been found at high altitudes. Some mammals are also found very high, including hares at around 19,600 ft (6000 m) in the Himalaya, and wild dogs above 18,000 ft (5500 m) on Kilimanjaro.

The highest zone of more or less continuous vegetation is alpine tundra, which is dominated by a mixture of low-growing grasses, herbs and shrubs, as well as lichens and mosses, whose proportion tends to increase at higher latitudes. Grasses often form tussocks, and herbs form rosettes or cushions that

Kilimanjaro (19,331 ft, 5892 m)
Africa's highest mountain.
Its ecosystems range from savannah
grassland on the plains at 4000 to
4600 ft (1200 to 1400 m) through
tropical forest, montane, and tundra
zones up to permanent glaciers.

Machhapochare, Nepal (22,950 ft, 6997 m) rises above the forests and irrigated rice terraces of the Middle Hills, more than 15,000 feet below.

retain moisture, provide protection of buds and flowering stalks, and enable insects to survive. The small size and rounded shapes of the herbs and shrubs means they can survive in harsh conditions, creating a microclimate that is far more suitable for growth and reproduction than only an inch or two higher. Rounded forms also offer less resistance to the wind and do not hold the snow. In high mountains, protection from high levels of ultraviolet radiation is ensured by the angle at which leaves grow, waxy surfaces and, sometimes, by hairs on the leaves. But what we see is only a small part of the total

vegetation: most of the biomass of these plants is below the soil surface in their roots, anchoring the plants and absorbing water and nutrients. Within the tundra zone, at altitudes up to 13,800 ft (4200 m) in the mountains of East Africa and the Andes – where the vegetation is known as *páramo* – are larger plants that rise a few feet above the rest of the vegetation: *Espeletia* in the Andes, and *Senecio* and *Lobelia* in East Africa. These strange-looking plants have a stem surmounted by a large rosette of fat leaves that opens during the day and closes at night, and are specially adapted to these high altitudes by having a tank root system that stores water.

Alpine meadows in the Alpachtal, Austria.

Traditional mountain societies use alpine tundra and pastures mainly for grazing animals. Depending on the species and the climate, the animals may live high on the mountains year-round, or descend to the valleys during the winter. Sometimes, they may be transported significant distances to lowland pastures. These traditional patterns generally persist in developing countries, but in Europe, the number of grazing animals tends to reflect the availability of subsidies from national governments or the European Union. When subsidies are high, numbers of animals – mainly sheep – increase, because subsidies bring in a higher income than meat or wool. The availability of low-cost grazing leases on government land is also a key factor setting the numbers of animals grazing in the mountains of the western United States.

Another major change has been the expansion of tourism. At one time, the only buildings found high in the mountains of Europe were small chalets or cabins; in other parts of the world, tents were used. Since the mid-twentieth century, this pattern has changed significantly in many mountain areas as their alpine zone has become the focus of tourism. In the Alps, the Andes, the Himalaya, and elsewhere, resorts have been built far above the long-established settlements, for skiing in winter and a variety of recreational activities in summer. Even higher in the mountains, mountaineering clubs have built huts, used as bases for mountaineering and skiing; this phenomenon began in the mid-nineteenth century in the Alps, and has since spread to mountain ranges all around the world as interest in these activities has grown, and accessibility has improved.

Pyramid Mountain, Jasper National Park, Canada.

Mountain Forests

One of the most obvious boundaries in mountain landscapes is the treeline, which marks the upper limit of continuous forests. There have been many theories about the reasons for this upper edge, which is sometimes abrupt, but often quite ragged. Recent research in mountain ranges around the world suggests that the natural treeline, both on mountains and at the edge of the boreal forests in higher latitudes, depends on the average temperature of tree roots; at temperatures below about 43°F (6.5°C) they cannot function properly. Trees growing close together prevent solar radiation from reaching the ground, resulting in lower soil temperatures, whereas trees growing apart from one another can survive because their roots grow into the warmer soils under surrounding tundra or grassland. Consequently, the natural treeline is determined by the interaction between the prevailing climate and the density of the trees. However, many treelines around the world are not natural; they have been lowered by people cutting trees for wood, or expanding the area of summer pastures, often by burning. Animals which graze on new or regenerating trees continue the process, maintaining the treeline at a lower than natural altitude.

The highest forests on Earth are composed of *Polylepis* trees, which grow in the Peruvian Andes just below the permanent snow line at 16,400 ft (5000 m). It is estimated that only 2 per cent of the original cover of these forests remains, after 10,000 years of burning, grazing by domestic animals, and climate change. In many other tropical mountains, the highest trees are of the *Erica* family, which also includes the heaths, rhododendrons, and blueberries, low shrubs that grow above the treeline in the temperate mountains of the northern hemisphere. In these mountains, the highest trees are typically conifers, especially pines, spruces, and firs. In some places, larches are present, being especially visible when their needles turn gold in the autumn. These forests also have rich populations of mosses and lichens. In other parts of the northern hemisphere, where there are few competing conifers – including Scandinavia, eastern Asia, and parts of the Himalaya – deciduous trees dominate at the treeline: mainly birches, but also alders, aspens, and beeches. In many of the temperate parts of the southern hemisphere, evergreen beeches (*Nothofagus*) are dominant in the upper forests, though there are exceptions, for instance in Australia, where snow gums form the treeline.

Forests cover over 3½ million square miles (9 million sq km) of the world's mountains: just over a quarter of the global area which is covered by forests. Nearly half of the area of mountain forest consists of coniferous species, particularly in the middle and higher latitudes of the northern hemisphere; another 773,000 sq miles (2 million sq km) are moist tropical forests. Mountain forests are composed of a huge number of both tree and other species, which are used in many different ways. The extraction of timber can be at any scale, from the removal of individual trees for construction and other purposes, to vast clear-cutting operations, such as those of western Canada, Siberia, and Southeast Asia, to provide timber and pulp for markets that are often far away. Mountain wood is also a vital resource for mountain people around the world, as it is the primary source of energy for most of them, and also for people living in nearby urban areas, either as fuelwood or as charcoal.

Trees are not the only components of mountain forests that are important for local economies.

In most of the mountains of Europe, the availability of cheap wood from other sources, such as Scandinavia and the lowlands of Central Europe, combined with the high costs of extraction in steep areas, means that mountain trees are barely, if at all, worth harvesting. However, other species are, particularly mushrooms, which often have a far greater value than the trees of the same forest, and cannot grow outside the forest. On average, a family in the Val di Taro of Parma, Italy, earns $2000 a year from *Boletus* mushrooms. High-value mushrooms are also harvested in other parts of Europe, western North America, and Asia. Medicinal herbs from mountain forests have a high value too, and the many animals living in forests can be valuable as sources of food, and for commercial hunting.

Mountain forests have many other significant, but less-easily measured values to both mountain and other people. The importance of the protection they provide against natural hazards, such as avalanches and rockfalls, has been recognized for centuries in some parts of the world; the first local regulations against cutting forests in the Swiss Alps in order to protect settlements were agreed in the thirteenth century. This protective function has become ever more important as the number of people not only living in, but also visiting and traveling through, mountain areas has increased dramatically since the mid twentieth century, largely due to the huge growth in mountain tourism; mountain forests play an important role in this as the setting for many sports and as a key element in the mountain landscape.

The importance of the protective function of mountain forests is increasingly recognized, but this awareness sometimes comes rather late. The first regulations in the Swiss Alps came after so-called 'natural hazards' had devastated settlements and crops — a pattern that recurred at ever-increasing scales, in spite of local regulations, until the first federal forestry law was passed in 1873. By then, the treeline had been lowered, on average, by 650 to 1000 ft (200 to 300 m). The law led to widespread replanting and regulation of forest activities. In the United States, vast areas of mountain forests were removed in the nineteenth century by logging and burning, leading to the first federal law in 1890. The forests have since recovered, and the area of mountain forests in Europe and North America has been increasing for some decades. Yet in tropical countries, both large-scale logging and cutting for shifting cultivation continue — and so do the disasters, which may be called 'natural', but are often largely human-induced. The greatest rate of deforestation of any type of forest is in tropical upland forests: 1.1 per cent of the total area a year.

The Himalayan foothills of eastern Nepal, during the monsoon.

*Terraces allow mountain people
to cultivate very steep slopes
with relatively little soil erosion,
as here in Nepal.*

Valley Floors and Agriculture

In a mountain valley which is uninhabited, or visited only by hikers, there will probably be marshes, meadows, forests, and thickets of shrubs. Many such valleys still exist, even in the relatively densely settled mountains of Europe, Asia, and Latin America, particularly in national parks and nature reserves. However, flat valley land is the best for agriculture and settlement, and many mountain valleys have been greatly changed by the clearance of vegetation, plowing, and the construction of buildings and transport infrastructure. As settlements grow, often taking up valuable agricultural land, their population densities may approach those of urban areas, leading to increased demands for food. For these reasons, and also because valley sides often provide the best conditions for cultivating certain crops, they are also brought into agricultural production. On steep slopes, terraces are often built – one of the most characteristic features of mountain landscapes found around the world. The construction and maintenance of terraces is very time-consuming, consequently, one measure of the pressure for agricultural production is the extent to which existing terraces are used for this purpose. Where demands are very high, all terraces are in use, with irrigation wherever possible to maximize yields; and there is further expansion of cultivation onto steeper and steeper slopes, with or without terracing. When demands decrease, the terraces which are on the steepest land and furthest away from settlements are used for grazing, or gradually allowed to return to shrub and forest. Once populations really start to decline, even terraces close to villages revert to forest, as in many of the mountains of Greece and Spain, where there may be large areas of uninhabited land by the middle of this century.

The decline of terraces can also be because men of working age migrate seasonally or for longer, to find work – this is one of the reasons behind the decline of the amazingly intricate terraces and irrigation systems in the mountains of Yemen, as very many men have left to work in the oil industry. Yet, even with emigration, some of the highest rural population densities in the world are found in tropical mountain areas, due both to local growth and to immigration by people from lowland areas, often leading to conflicts over land and other resources and the need to produce ever more food. One way to do this is to expand on to steeper slopes, shorten rotation periods, bring in more water (if it is available) through irrigation, and apply more fertilizers (if they can be afforded). However, increased pressure on the land often leads to the depletion of soil nutrients, soil erosion, and declining crop yields.

Another way to increase production and stabilize the land is through agroforestry, which is a way of creating ecosystems that are similar in structure to an area's natural ecosystems, but are composed as much as possible of species that can be used by people. One particularly good example of a traditional agroforestry system, operating at least since the seventeenth century, is the Chagga home gardens, which cover an area of around 297,000 acres (120, 000 ha) on the slopes of Kilimanjaro, and which produce bananas, beans, cardamom, coffee, onions, yams, and timber. Stingless bees in the home gardens produce honey that is five times more valuable than honey from ordinary bees because of its medicinal value. Most of the farmers also have land on the drier plains where they grow staple foods.

In the Himalaya and Andes, shade trees are widely used in coffee and tea plantations. Trees that bear fruit or have leaves that can be used as fodder for animals or as organic manure can be planted along terraces, also anchoring them. Leguminous species, such as beans, peanuts, and peas, can be planted, which not only provide food, but also increase soil fertility by absorbing nitrogen from the air. A considerable proportion of many crops are lost to pests, during harvesting, and by subsequent deterioration and loss to pests and animals. Better pest control, harvesting, and storage systems are also important issues to be considered with regard to increasing food security for mountain people.

Mountain Settlements

Although most mountain people live in rural areas, there are also major urban centers in the mountains, especially in tropical and sub-tropical areas, where mountains are often preferred as places to live. Many of the major cities in Central and South America are in the mountains, including Mexico City, one of the largest on earth. In Asia, there are also large cities within mountain areas, particularly in China, and very close to them, including two of the largest cities in the world, Tokyo and Jakarta. In Europe, there are a number of cities with populations over 100,000 within and immediately next to the Alps, and in other parts of the continent. Recent research, bringing together census data with satellite measurements of the light visible from the Earth at night, suggests that 1.48 billion people – 26 per cent of the world's population – live in or very near to mountain areas.

The town of Jasper, Jasper National Park, Alberta, Canada.

Centers of Diversity

I depend mainly on agriculture like most people in this village. We grow maize, sorghum and different vegetables. We also have wild vegetables to supplement our diet in case of severe drought. Basically, we are self sufficient in this village. The surrounding villages often come to our village for food and we sell them whatever surplus we have.

Makibinyane, a farmer from Molika-liko, Lesotho. *Mountain Voices (1997)*

Biodiversity Hotspots

Many people believe that tropical lowland rainforests are the parts of the Earth with the greatest biological diversity; yet tropical mountain forests actually have the greatest levels of biodiversity. For example, in Ecuador, 6500 sq miles (17,000 square km) of tropical mountain cloud forest contain 3411 plant species: 300 more species than in 27,000 sq miles (70,000 square km) of adjacent lowland rainforest. It is estimated that there are 7.5 times more moss species in the five tropical Andean countries than in the entire Amazon basin. The eastern Andes is one of the centers of greatest biodiversity, or 'biodiversity hotspots', on Earth. Most of the others are also in, or include, tropical mountains: the Atlantic forest of Brazil; northern Borneo; the eastern Himalaya-Yunnan region; and Papua New Guinea. Many secondary centers are in Mediterranean areas – which have the greatest number of tree species outside the tropics – as well as parts of the Rocky Mountains, and the subtropical arid mountains of Central Asia.

There are many reasons for these high levels of biodiversity. The steep altitudinal gradient provides a rich variety of habitats at all scales, and is a major reason why tropical mountain forests are more biodiverse than adjacent lowland forests. This variety brings together many environmental factors, including the gradient from dry upper to wet lower slopes with accumulated nutrients and debris; differences in soil depth and type; contrasts between sites, from very exposed to sheltered; and differences in the degree of disturbance, from events such as avalanches and landslides.

As mountain ranges have developed, species have been able to migrate along new pathways, exploiting ecological niches as they have opened up. Interruptions in mountain-building phases, subsequent erosion, and changes in climate – especially ice ages – have also isolated species, so that

Mount Semeru, Java, is the highest active volcano in Indonesia.

Many mountain landscapes are cultural landscapes, fashioned by people over centuries to provide a wide range of resources at different altitudes. In Hohe Tauern National Park, Austria (left), most of the forests remain, interspersed with agricultural land. In Nepal's Middle Hills (right), the landscape has been changed more radically to support much higher population densities through the cultivation of crops including millet, potatoes, and soya beans.

they have evolved in different ways. These two sets of factors are key reasons why mountain regions have such a high proportion of endemic species, restricted to just one mountain range – and sometimes to just one mountain. This applies not only to plants, but also to other species: for instance, about half the internationally recognized Endemic Bird Areas worldwide are in mountain areas, particularly tropical mountain forests.

A third set of factors leading to the high biodiversity of mountain ecosystems derives from human activities. While lowland areas have been cleared and cultivated for centuries, if not millennia, the steep slopes and less attractive conditions for agriculture found in mountains often meant that they were largely left alone. This is especially evident in Europe, where mountains can immediately be recognized on a map of biodiversity hotspots. The Alps, for example, host about 4500 vascular plant species: more than a third of the entire European flora, of which about 15 per cent are endemic.

Benefits of Biodiversity

In general, as one goes up a mountain, the number of species decreases. Alpine tundra and grassland habitats, and upper mountain forests, tend to have a high diversity of species, but a low diversity of genera or families. Lower forests have a high diversity of both genera and families. This rich natural biodiversity is of great value to people for many purposes. Alpine ecosystems provide medicinal plants as well as good grazing for animals; forests provide different types of wood, fruits, mushrooms, and medicinal and aromatic plants. An example of these riches is provided by Peru, a high-diversity and mountainous country, where 3140 out of 25,000 known vascular plant species are used by people, including 444 for wood and construction, 292 for agroforestry, 99 for fiber production, and others as cosmetics, narcotics, stimulants, dyes, and ornamentals.

Another example comes from the mountains of Nepal, where there are 45 species of exportable medicinal and aromatic plants in the alpine zone, 114 in the sub-alpine zone, 225 in the temperate zone, 340 in the sub-tropical zone, and 310 in the tropical zone. The availability of many of these species is ensured by careful tending and planting in ecosystems that might appear natural, but are actually partially cultivated. Yet only 100 of these species are currently harvested for commercial use. Annual harvests, mainly exported to India, may be 15 to 42,000 million tons a year, worth US$ 8.6 to

The Incas built the major ceremonial centre of Machu Picchu, Peru, in the mid fifteenth century, on a ridge surrounded by sacred mountains and almost encircled by the sacred Urubamba river. It was abandoned in the early sixteenth century in a period of civil war and disease. This World Heritage Site now attracts over 300,000 visitors a year.

26.7 million. However, the mountain farmers who collect the plants receive very little of this value, which is mainly taken by middlemen. Cultivation, processing, and marketing methods must be improved to ensure that local people derive a more equitable share of the value and benefits from such vital resources.

Another way in which the wider world has benefited from the genetic diversity of mountain areas is with respect to the crops that originate in these areas. The introduction of potatoes and maize from Latin America has made possible the maintenance and expansion of both mountain and lowland populations in many parts of the world. The mountains of the Near East are the home of the original precursors of wheat, and these original varieties are still important in the breeding of varieties of major food crops for specific characteristics, including resistance to disease. There are also many species that are not widely known but are adaptable and nutritious, such as many of the Latin American root and tuber crops. Individual farmers are key to preserving this variety: some Andean farmers plant over 300 varieties on their widely spaced plots, each with specific characteristics. Research centers, such as the International Potato Center (CIP) in Peru, are also important: CIP has collected and stores over 3500 varieties of potatoes and many other crop species. It now has an exchange program with other centers in the Himalaya and Africa, to determine the suitability of crops for other regions: a crucial need in a world with a growing population and changing climate.

Mountain Cultures

Mountain areas are centers of not only biological, but also cultural diversity, and these are often closely related. One measure of cultural diversity is in terms of the number of languages and dialects spoken in a region, and of differences between these languages and dialects and those of adjacent lowlands. For example, there are four official languages in Switzerland – French, German, Italian, and Romansch – and many dialects of each of these, between which mutual comprehension can be very difficult. In northern Pakistan, an area described as a 'giant ethnographic museum' because of its great cultural diversity, the 35,000 people of the Hunza valley, about a quarter of the area of Switzerland, speak four different languages belonging to three language families.

Most mountain people are parts of larger cultural groups, such as the French-, German-, and

Italian-speaking people of Switzerland. There are also groups who only or mainly live in the mountains, such as the speakers of the old Romance languages (such as Romansch) in the Alps, the majority Burosho people of Hunza, and larger groups in other parts of the world, such as the Amhars of Ethiopia, the Quechua of the Andes, and the Tibetans and Yi of China. Cultural diversity can also be recognized in differences in belief systems, building styles, clothing, crops, customs, and music.

Many of the reasons for this cultural diversity may be similar to those leading to high levels of biological diversity, such as isolation, refuge from dominant cultures, use of a diversity of ecosystems, and existence far from centers of power. However, their interactions are much more complex: do people retain their specific characteristics because of their relative isolation or peripherality, or because they want to do so, in order to retain their identity – or even to benefit by presenting a particular image to tourists? One thing is certain: no mountain group has ever been completely cut off from other people; even in the least accessible mountain areas, people have traveled out and in, usually to trade goods. Mountain people are increasingly being integrated into the wider world, as young people leave and return, radio and videos become available, and tourists arrive.

The distinct clothing, festivals, handicrafts, and agricultural products of mountain people are among the great tourist attractions but they often lose some of their original meaning, for example if the timing of festivals is changed to coincide with – or avoid – the main tourist season, if dances that were formerly performed only once a year are performed daily during the tourist season, or if festive foods are prepared specially for tourists. In both Thailand and the Andes, traditional clothing is sold to tourists in search of authentic souvenirs – yet, conversely, the expectation of tourists led the villagers of Hermagor in the Austrian Alps to design a new local costume! The demand for souvenirs and local food and drink can lead to a renaissance of traditional crafts, though often the quality may not be as high as previously; and the money earned from tourism can be used to invest in temples and monasteries, as has happened in the Sherpa area of Nepal.

Mountain cultures, like their ecosystems, crops, and grazing animals, are often very distinct, but as mountain people become increasingly integrated into the wider world, their cultures will continue to change. The great challenge is to ensure that change is not so abrupt that they lose their identities and self-confidence and become just actors on the tourist stage, or indistinguishable from everyone else.

Mountain areas are global centres of cultural diversity, often seen in traditional ways of life and dress.
The dancers of Papua New Guinea's Huli tribe (left) are wearing ceremonial clothing, with wigs made of their own hair, feathers, and flowers. The girls (right) are from the Lisu people, who live in northern Thailand as well as Myanmar and China.

53

Places of Attraction, Challenge, and Renewal

Thousands of tired, nerve-shaken, over-civilised people are beginning to find out that going to the mountains is going home.
John Muir, 'The wild parks and forest reservations of the West', *Atlantic Monthly 81* (1898)

Only two centuries ago, most people in western cultures regarded mountains as unattractive, dangerous places. Since then, their beauty and challenge have come to be recognized as key attractions. As more and more of us come to live in cities, mountains are becoming increasingly important as places which raise our spirits, giving us opportunities to escape from everyday life and experience different environments, cultures, and sports.

Ever since people have lived in the mountains, some have climbed them to search for better grazing or lost animals – and probably for fun. It was not until the nineteenth century that any number of Europeans and North Americans began to travel to the mountains for pleasure, both to look at the scenery and to climb on their rock faces and glaciers to the summits. Mountaineering is now a world-wide sport, with millions climbing in countries around the world. Mountaineers have often been the first foreign visitors to mountain areas, and the images they have brought home, through books, films, and television, have been a major reason for the expansion of mountain tourism, one of the greatest causes of change in mountain societies. Nearer home, the millions of people who visit the mountains every weekend to hike, bike, climb, or just look at the scenery, have had major influences on the landscapes and economies of the mountain areas that are easily reached from urban centers.

Sacred Mountains

The relatively new role of mountains as places which attract tourists has deep roots in many cultures. For millennia, the scale and dramatic weather associated with mountains – as well as recognition of their role as 'water towers' – has given them great spiritual significance. Almost every religion identifies individual mountains, or special places within them, as sacred: places to which people come for spiritual renewal and worship. Mountains may be the home of one or more gods, or even – as in New Zealand and Bolivia – of ancestors themselves. Mount Kailas, in remote western Tibet, is the most sacred

Northeast face of Mount Kailas, Tibet, at dawn.

mountain for both Buddhists and Hindus – almost a billion people. Every year, thousands of pilgrims make a circuit around the mountain, many on their knees the whole way.

Many mountains are visited mainly for religious reasons, such as Ausangate in Peru, Croagh Patrick in Ireland, and Mount Athos in Greece. Other sacred mountains are visited by vast numbers of people, whose motives for visiting may be only slightly religious at most. These include Mount Fuji in Japan, visited by over a million people a year; T'ai Shan in China, now with a cable car to its summit; and Jebel Musa or Mount Sinai in Egypt, a destination for day trips from the Red Sea coast.

While Mount Kailas is the most sacred mountain for Hindus, relatively few ever go there. Yet, on the other side of the Himalaya, in the mountain states of Himachal Pradesh and Uttar Pradesh in India, nearly 10 million pilgrims a year enter Dev Bhumi, the land of the gods, in the Garhwal region, which has many temples and shrines. Major sites of pilgrimage have existed in this region since at least the beginning of Hindu civilization. Nearly half of the domestic product of Uttar Pradesh comes from tourism; 60 per cent of the tourists are pilgrims, almost all from India.

One major site is Badrinath, attracting nearly half a million people a year: a three-fold increase since the 1970s. One reason is improved accessibility along the road built into the area after the 1962 war between India and China. The road increased the number of tourists, but meant that most of the people who had provided food, accommodation, and transport along the old pilgrim route lost these additional sources of income. Today, immigrants from other parts of India run more than half of the businesses in Badrinath, and they have put pressure on the local government to lengthen the period that the temples are open – a proposal against which local people have strong feelings. The massive influx of tourists also led to major problems of sanitation and waste management, and the loss of most of the forests. This trend is now being turned around by a partnership between local scientists and priests. Pilgrims are encouraged to plant trees, through sermons which emphasize the importance of the trees in minimizing floods and landslides – as well as the prospect of gaining additional blessings by helping to restore an ancient sacred forest.

Harnessing the World's Largest Industry

The example of Badrinath shows many of the problems associated with the development of tourism

Mount Fuji (12,388 ft, 3777 m) is the abode of a Buddhist deity.
The Buddhist word for its summit, 'zenjo', means a flawless state of perfect concentration. During the official
climbing season, over a million people climb the mountain – but few are now religious pilgrims.

Mont Blanc, the highest peak in western Europe (left, 15,771 ft, 4807 m) was the first major Alpine peak to be climbed, in 1785. Today, its glaciers, rock faces, and ridges are one of the major destinations for climbers from around the world, and millions of other tourists who come to enjoy the scenery.

in mountain areas, but also that innovative solutions can be found. Tourism is the world's largest industry, and it is estimated that 15 to 20 per cent of this, or $540 to 720 billion a year, is associated with travel to mountain areas. The rapid growth of tourism as a key element in mountain economies is due to several factors, such as the growing numbers living in urban areas, and increases in discretionary time, income and mobility, allied to our need to 'escape' from urban areas and the remarkable increase in accessibility of mountain areas, either via roads that were built for other reasons, such as the extraction of natural resources or military purposes; or that were built mainly for this purpose, for instance to ski areas. Railways opened up a number of valleys in the Alps and Indian Himalaya to tourism in the nineteenth century and, in the last century, the 'bullet train' system linked Japan's major cities to the mountains. The real costs of international air travel have decreased and, within countries, small airlines and helicopters make it possible to reach almost any mountain area.

Tourism has become a major force of change in mountain areas around the world, often regarded by both governments and mountain communities as vital for economic development and survival. Yet its benefits are usually spread very unevenly, at whatever scale you look, from the national to the individual. At the scale of communities, the apparent benefits of tourism in terms of maintaining populations are shown by statistics from the Alps, which account for 7 to 10 per cent of the annual global tourism turnover. Forty per cent of communes have no tourism, and only 10 per cent depend predominantly on tourism. Generally, the former are losing population, while the latter have stable or growing populations. On an individual level, a small number of people may get rich from tourism; they are often immigrants or others with access to outside capital. Yet their success may depend on the majority who work for relatively low wages, and may not even be able to find housing where they work. For example, people working at some ski resorts in the Rocky Mountains of Colorado, in the United States, have to commute for nearly an hour to work, sometimes over high mountain passes.

Mountain tourism is a massive and complex interaction of visitors and mountain people involved in a vast array of sub-sectors, from pilgrimage to mass tourism, cultural tourism, ecotourism, health tourism, and an incredible variety of sport tourism. Each involves a different clientele in a highly competitive, unpredictable, and increasingly global market. Each type of visitor demands a specific range of services and facilities. Changes in fashion often mean that investments made to attract one type of tourist must be supplemented by additional investments in new facilities to encourage repeat visits or attract new types of tourists. As well, most types of mountain tourism are highly seasonal. Skiing is only possible in cold temperatures, relying on natural and, increasingly, expensive man-made snow. Many types of ecotourism are related to the annual cycles of particular plants and animals. People coming for hiking holidays prefer the seasons with the least rain and annoying insects.

People who work in the mountain tourism industry often find that services and facilities suitable for tourists in one season are not appreciated by those who come at other times. For instance, visitors who come to enjoy the summer landscape do not like to see ski lifts and other evidence of winter activities. Visitors have a clear image of the landscapes they expect to see; mountain people need to maintain these so that tourists continue to come, which means that land uses such as agriculture and forestry contribute directly to the tourists' experience. Marketing is essential to attract tourists; but the image that they leave with, and communicate to others, may be just as important in ensuring future visits.

While mountain tourism can be an important source of income in both industrialized and developing countries, its long-term unpredictability and seasonality mean that it should always be integrated with other economic sectors. People working in tourism should maintain their opportunities for employment and earnings in other ways, whether agriculture, forestry, industry, handicrafts, tele-working, or commuting in the off-season. To prosper in this competitive industry, those involved in tourism in every mountain community need to develop a unique image based on their local environment and culture – and to invest income from tourism in these to ensure a future that is sustainable, even if numbers of tourists decline. Rather than focussing just on tourism, governments and development agencies should also recognize its complex links with other sectors when they develop projects and implement policies, and provide the means for training in skills necessary both in tourism and for other employment opportunities. And tourists should be aware of how they can contribute to mountain environments and people, choosing to travel with companies that employ local people, staying in locally owned accommodation, eating and drinking local produce, insisting on the use of renewable energy sources – such as kerosene rather than firewood – and buying locally made souvenirs.

Midi d'Ossau, Pyrenees, France.

Safeguarding Our Mountain Heritage

Mountain peoples, in their vertical archipelagos of human and natural variety, have become the guardians of irreplaceable global assets.

Derek Denniston, *High Priorities: Conserving mountain ecosystems and cultures.* Worldwatch Institute, Washington DC, 1995

Around the world, the most remarkable scenery, landscapes, cultural sites, and biologically diverse ecosystems are protected under the 1972 World Heritage Convention as World Heritage Sites of 'outstanding universal value'. The global importance of the natural and cultural heritage of mountains is shown by the fact that almost two-thirds of the natural and 'mixed' (with both natural and cultural importance) sites, and one sixth of the cultural sites are in mountains. Designation of an area or cultural site as a World Heritage Site is the highest accolade that can be bestowed by the global community. The importance of many mountain landscapes and ecosystems is also recognized by their designation by governments as national parks, nature reserves, and other types of protected area. In 1997, it was estimated that a third of the global area protected by such national designations was in mountains. This was certainly an underestimate, as it included only protected areas with a minimum size of 25,000 acres (10,000 ha) and a minimum altitudinal range of 5000 ft (1500 m). For instance, no protected mountain areas of Britain were included, as Britain's highest point, Ben Nevis, is only 4406 ft (1343 m) high.

The first national park in the world, Yellowstone in the Rocky Mountains of the United States, was designated in 1872. Many of the other early national parks established around the world, including Banff in Canada and Tongariro in New Zealand, are also in mountain areas. For about the first century of the protected area movement, 'protection' often meant that local people were largely excluded from national parks. In some places, including the Swiss National Park, the Gorge of Samaria on the island of Crete, and very often in developing countries, local people were removed when parks were established, and their villages were left to decay, or were destroyed.

Since the 1980s, there has been increasing recognition that protected areas cannot be managed as 'islands' separate from the surrounding landscape; wider, regional approaches are needed. This means

Moraine Lake, Banff National Park, Alberta, Canada.

that these special places should no longer just be protected by government agencies or organizations focussing on nature conservation. To ensure the long-term survival of the species, ecosystems, and landscapes within a particular area, local people – and other 'stakeholders' such as local governments and non-governmental organizations (NGOs) – must be fully involved. Just as tourism should be considered within the broader context of economic development and environmental management, so should the management of protected areas and conservation as a whole, recognizing that people, wildlife, fires, diseases, and air pollution regularly cross administrative and ecological boundaries.

Partnerships for Conservation

One characteristic of traditional mountain societies is cooperation, and so the development of partnerships involving different players in the common interest has deep-rooted origins. In many mountain regions around the world, such cooperative structures have decayed, or been marginalized, as regional and national governments have taken away many of the rights and responsibilities of mountain people over their resources, often giving them to government agencies staffed by people – often from far away – with scientific and technical training. The knowledge of local people, based on centuries of experience, has been ignored, and they have been excluded from using the resources of protected areas. In some cases, this has meant that some of the qualities for which an area was originally designated could be lost. For example, after the national park was declared in the Tatra Mountains of Poland, local people were not allowed to graze their animals there. One result was that some rare plant species on formerly grazed meadows began to decline in number because they were being shaded out by taller plants that had previously been grazed down. Fortunately, this was recognized, and farmers were invited to bring their animals back into the park – and the populations of the rare plants recovered. Similarly, in the Vanoise National Park in the French Alps, local people are now paid to mow species-rich meadows, replacing grazing animals. In such ways, conservation is linked to the continuation of local practices and employment possibilities. These two examples underline the fact that many mountain landscapes around the world are not entirely natural; they are cultural landscapes, shaped by centuries of human action.

For government agencies, the shift from 'expert management' to management in partnership is

Much of Peru's Cordillera Blanca, the world's highest tropical mountain range, has been protected within Huascaran National Park, which includes many peaks over 20,000 ft (6000 m). It is also a natural World Heritage Site and Biosphere Reserve.

usually not easy; it requires giving away power and authority, and recognizing that local people have complementary knowledge and rights to be involved in the management of their region. The development of such partnerships requires careful planning and negotiation. One region where this has been a long, difficult process is around Yellowstone, where the government agencies responsible for the national park and most of the surrounding forests published a vision for cooperative management of the wider region in 1987. Many saw this as a great advance, as it brought together

Milford Sound, Fiordland National Park, New Zealand.

two agencies that had not always cooperated, and was based on the latest scientific approaches. But, it took little account of local interests and knowledge. As local people were not included early enough, it took many more years to build their trust in order to move forward towards regional management.

Throughout the process of developing a partnership, and working within it, governments and their agencies need to consider both local goals and values, and those of the wider society. In mountains, local needs may relate to employment, subsistence, and sources of energy; broader societal needs may relate to the availability of recreation and tourism opportunities, the conservation of biodiversity, and the provision of freshwater. A key element in building partnerships is to ensure that all partners have access to the same information. For instance, the indigenous people of the Sierra Nevada de Santa Marta of Colombia were provided with computer-based geographic information systems, and trained to use them, so that they could map their landscape and evaluate the implications of different potential land uses. Also, when the dominant national language is different from the languages of mountain people, documents should be available in these languages, in forms that can be easily understood.

The expertise of local people must also be acknowledged; when programs to monitor the success of nature conservation measures are being developed, it may be much more appropriate to use this expertise, as has been done in Uganda's Rwenzori mountains, rather than expensive expatriate consultants or scientific equipment. In such ways, cultural heritage is preserved, and people gain employment through being stewards of their own landscape.

There can also be other incentives for local people to work in the long-term interests of conservation. For instance, poachers can be encouraged to become game wardens and guides, receiving a salary for protecting animals that tourists come to photograph, rather than shoot. Such an approach now protects some of the snow leopards in the mountains of northern Pakistan. Entrance fees can also be reinvested in local community facilities, as for instance in Bwindi national park in Uganda, where the park authorities have also provided local people with bamboo rhizomes to plant on their farms. More and more, conservation authorities and mountain people are coming together to find ways of cooperation that benefit all parties, as well as the environments and species they cherish.

Cooperating in an Uncertain World

Ensuring that the unique heritage of mountain areas is preserved and nurtured, and that mountain areas continue to provide benefits far outside their boundaries, requires ways of thinking that may seem new, but are often rooted in old traditions in regions where cooperation has been essential to surviving in an unpredictable environment. The future of the environments and the people of many of the world's mountains are ever more uncertain. Mountain people are increasingly and inevitably involved in regional and global economies; but they also include a significant proportion of the world's 800 million undernourished people. Because of poverty, high ethnic diversity, and external political and economic forces, a disproportionate number of armed conflicts occur in the mountains: in 1999, 23 of the 27 major armed conflicts in the world were being fought in mountain regions. Long-established ways of life and resource management, well adapted to often difficult circumstances, are also endangered by emigration, the arrival of tourists and new forms of electronic communications.

Yet returning migrants, as well as those who move to the mountains because they perceive them as attractive places to live, can bring new ideas and money to invest, and tourists, outdoor recreation, and telecommunications can provide opportunities for making a living in many mountain areas.

Our global awareness also allows us to realize that the uncertainty of mountain areas may be increased by climate change. The accelerated melting of almost all of the world's glaciers provides some of the best evidence for this process, which is likely to result in greater numbers of extreme events – such as hurricanes, windstorms, and floods – as experienced recently in mountain areas including Central America, the Giant Mountains, and the Alps. Fortunately, in this uncertain world, more and more partnerships are being developed, from cooperative groups charting the future of individual mountain valleys; to regional partnerships which bring together conservation, tourism, forestry, and other interests; to the Mountain Forum, a world-wide network of people sharing information in the interests of sustainable mountain development, which has grown out of the mountain chapter of 'Agenda 21'. One of the best mountain partnerships that has emerged in recent years is the Panos Institute's 'Mountain Voices' project in which local people in many mountain regions have been trained to interview people in their local communities. The interviews, which are translated and archived on a website, provide a comprehensive picture of highland societies. Ram, a farmer from Sindupalchok in Nepal, summarizes some of the challenges and solutions for everyone concerned with the mountains:

'In our village, sometimes neighbours become sick and are not able to help in time, at that time villagers will help them…. Sometimes I have to face more problems myself. At that time I need help, I would not be able to do my work by myself. Then my neighbours will decide to help me. If they do not help me I will be ruined. So this is a good system. When the time comes I will also help them. We have a good cooperation between us.'

Such partnerships are essential not only in mountain villages, but to bring together all those who recognize how important mountains are. The prosperity of mountain people and the environments on which they, and most of us, depend requires not only cooperation, but adaptation, mutual understanding, and willingness to learn from others, recognizing that 'we are all mountain people'.

Everest and Nuptse at sunset from Kala Pattar, Khumbu, Nepal.

Recommended Reading, Information and Web Links

There are a huge number of books on every aspect of mountains. The best global overview of mountain issues is Messerli, B. and J.D. Ives (eds.) *Mountains of the World: A global priority.* Parthenon, 1997. The best textbook, though outdated in some repects, is Price, L.W., *Mountains and man: A study of process and environment.* University of California Press, 1981. It will be republished as Friend, D.A., Byers, A.C. and Price, L.W., *Mountains and People,* Berkeley, University of California Press, 2004. Another book giving a very broad overview of mountain issues, with examples from around the world, is Price, M.F. and Butt, N (eds.) *Forests in Sustainable Mountain Development.* CABI Publishing, 2000.

The quarterly journal *Mountain Research and Development* (www.mrd-journal.org) is a primary source of new research. Other good websites, with many links, are the Mountain Forum (www.mtnforum.org), the International Year of Mountains (www.mountains2002.org), the International Center for Integrated Mountain Development (ICIMOD) (www.icimod.org.sg/), the International Potato Center (CIP) (www.cipotato.org/), Mountain Voices (www.mountainvoices.org/) and the mountains part of the People and Planet website (www.peopleandplanet.net/doc.php?id=966&section=11).

IUCN – The World Conservation Union. Founded in 1948, The World Conservation Union brings together States, government agencies and a diverse range of non-governmental organizations in a unique world partnership: over 980 members in all, spread across some 140 countries.

As a Union, IUCN seeks to influence, encourage and assist societies throughout the world to conserve the integrity and diversity of nature and to ensure that any use of natural resources is equitable and ecologically sustainable.

The World Conservation Union builds on the strengths of its members, networks and partners to enhance their capacity and to support global alliances to safeguard natural resources at local, regional and global levels.

Buachaille Etive Mòr, Glencoe, Scotland, U.K.

Index

*Entries in **bold** indicate pictures*

accessibility 38, 56, 59
Adriatic Sea 33
Afghanistan 26
agriculture 13, 30, 31, 43, 47, 49, 60
agroforestry 44, 49
Alps 7, **8**, 13, 17, 22, 25, 26, 34, 38, 40, 44, 49, 52, **58**, 59, 64, 68
Amu Darya River 31
Andes 13, **16**, 17, **28**, 29, 34, 37, 38, 39, 44, 47, 52
Antarctica **12**
Appalachians 13, 17, 20
Argentina **32**, 33
Atacama Desert, Chile **28**, 29
Ausangate, Peru 56
Australian Alps 13
Austria 13, **37**, **48**, 52
avalanches 20, 40, 47
Badrinath, India 56
Balkans 34
Ben Nevis, Scotland, UK **1**, 63
biodiversity 47, 49, 66
Bolivia 17, 55
Borneo 47
Brazil 13, 30, 47
   Atlantic Highlands 13
Buachaille Etive Mòr, Scotland, UK **70**
Bulgaria 8, 13
Canada 7, 8, **9**, 30, **38**, 39, 44, **45**, **62**, 63, 72
Central Asia 31, 47
China 11, 13, 22, 44, 52, 53, 55, 56
cities 44, 55, 59
climate change 11, 20, 26, 31, 68
Columbia River 30

Cordillera Blanca, Peru **65**
Cotopaxi, Ecuador **24**
Croagh Patrick, Ireland 56
crops 26, 30, 34, 40, 43, 51
Cuernos del Paine, Chile **10**
cultural diversity 51, 52, 53
dams 30
deforestation 11, 33, 40
Earth Summit 11
earthquakes 20, 22
East Africa 25, 37
ecotourism 59, 60
emigration 67
endemic species 49
erosion 18, 20, 42, 43, 47
Ethiopia 25, 30, 52
Euphrates River 31
European Union 37
farmers 44, 51, 68
floods 20, 22, 30, 56, 68
forests 8, 29, 30, 33, 34, 36, 38, 39, 40, 47, 48, 49, 56, 66, 71
France 17, 60
Ganges River 31
genetic diversity 51
Germany, Black Forest 17
Giant Mountains 17, 68
Gilkey Glacier, Alaska, U.S.A. **23**
glaciers 7, 8, 18, 20, 22, **23**, 24, 26, 33, 55, 58, 68
glacial lake outburst floods (GLOFs) 22
grazing 30, 33, 37, 38, 39, 43, 49, 55, 64
Greece 43, 56
handicrafts 29, 52, 60
Hawaii, USA 17, 18, **19**
Himalaya 17, 29, 34, 38, 39, **41**, 44, 47, 51, 56, 59

Hindu Kush 17, 29
Huascaran, Peru 22, 65
Hunza, Pakistan 30, 51, 52
hydro-electricity 25, 29, 30, 31
hydropower 22, 30
Iceland 18
immigration 43
International Year of Mountains 11, 71
Iran 26
irrigation 25, 26, 30, 31
Italy 17, 40
Jebel Musa 56
Jordan River 31
Kazakhstan 31
Kilimanjaro 34, **35**, 44
Kyrgyzstan 11, 31
landslides 20, 47, 56
languages 51, 52, 67
laws 11, 30, 40
Macchu Picchu, Peru **50**
maize 47, 51
Matterhorn **8**
Mauna Loa, Hawaii 17
medicinal plants 49
Mediterranean 33, 47
Middle East 26
Mont Blanc **58**
moraines 20
Mount Athos 56
Mount Etna 17
Mount Everest 3, **69**
Mount Fuji 18, 56, **57**
Mount Kailas **54**, 55, 56
Mount Pinatubo 17, 22
Mount Rundle **9**
Mount Semeru **46**
Mount Sinai 56
Mount St. Helens 17, 20, **21**
mountains

definitions 13
formation 17
Mountain Agenda 11
Mountain Forum 68, 71
Mountain Voices 47, 68, 71
mountaineering 38, 55
mushrooms 40, 49
national parks 4, 8, 33, **38**, 44, **45**, **48**, **62**, 63, 64, **65**, **66**, 67
natural hazards 20, 22, 40
Nepal 22, 26, **27**, 30, **36**, **41**, **42**, **48**, 49, 52, 68, **69**
   Machhapochare **36**
New Zealand 55, 63, **66**
Nile River 25, 31
non-governmental organizations (NGOs) 64, 71
North Africa 17, 26
North America 13, 17, 40
Norway 30
orogenies 17
Papua New Guinea 34, 53
páramo 37
partnership 56, 64, 66, 68, 71
pastoralists 34
Peru 22, 39, 49, **50**, 51, 56, **65**
pilgrimage 56, 59
potatoes 48, 51
poverty 67
Pyrenees **61**
railways 20, 33, 59
rain 26, 29, 34, 60
recreation 25, 38, 66, 71
rockfalls 20, 40
Rocky Mountains **4**, 7, **9**, **38**, 47, 59, **62**, 63, 66, 72
Rwenzori Mountains, Uganda 67
Scandinavia 17, 39, 40

Scottish Highlands UK 13, 17, **70**
Sierra Nevada de Santa Maria, Colombia 66
skiing 38, 59, 60
Slovenia 13
Snowdon range, Wales, UK **6**
Spain 13, 43
Switzerland 7, 30, 51, 63
Syr Darya River 31
T'ai Shan, China 56
Tatra Mountains, Poland 64
tele-working 60
terraces 20, 36, 42, 43, 44
Thailand 52, 53
Tibet 22, 55
tourism 29, 38, 40, 52, 55, 56, 59, 60, 64, 66, 68
tourists 20, 29, 52, 60
transhumance 33
treeline 38, 39, 40
tropical forest 35, 39
tundra 33, 34, 35, 37, 38, 49
United States **4**, 7, 17, 22, 30, 38, 40, 59, 63, 72
Urals 13
Uzbekistan 31
volcanoes 18, **24**, 27, **46**
water 11, 18, 20, 22, 25, 26, 29, 30, 31, 34, 37, 43, 55
   conflicts 31, 43, 67
   towers 5, 25, 31, 55
weathering 18, 20
wheat 51
World Heritage Site 7, 50, 63, 65
Yellowstone 63, 66
Yemen 29, 43

Biographical Note

Martin F. Price is the Director of the Centre for Mountain Studies at Perth College, within the UHI Millennium Institute in the Highlands and Islands of Scotland. He has worked on mountain issues since the late 1970s, in the Rocky Mountains of Canada and the U.S.A., in many European mountain ranges, and also at the global scale. This has included involvement in the process associated with the formulation, implementation and review of the outcomes of the mountain chapter of 'Agenda 21'. He has been a consultant to many international organizations, including IUCN - The World Conservation Union, FAO, UNEP, UNESCO, and the European Commission, on both mountain topics and various aspects of global change. He has written and edited several books and over 100 reports and articles for both scientific and lay audiences.